A Soldier's Story:

Prison Life and Other Incidents
in the War of 1861 - 1865
Elmira Prison Camp

by Miles O. Sherrill

1904

A Soldier's Story: Prison Life and Other Incidents in the War of 1861-1865 - Elmira Prison Camp

by Miles O. Sherrill, originally published in 1904, with added information by New York History Review, 2016. Photographs by Allen C. Smith.

For information on getting permission for reprints and excerpts, contact us through our website:
www.NewYorkHistoryReview.com

First Edition
ISBN: 978-0-9965353-7-3

Printed in the United States of America.

To all Confederate Soldier descendants.....

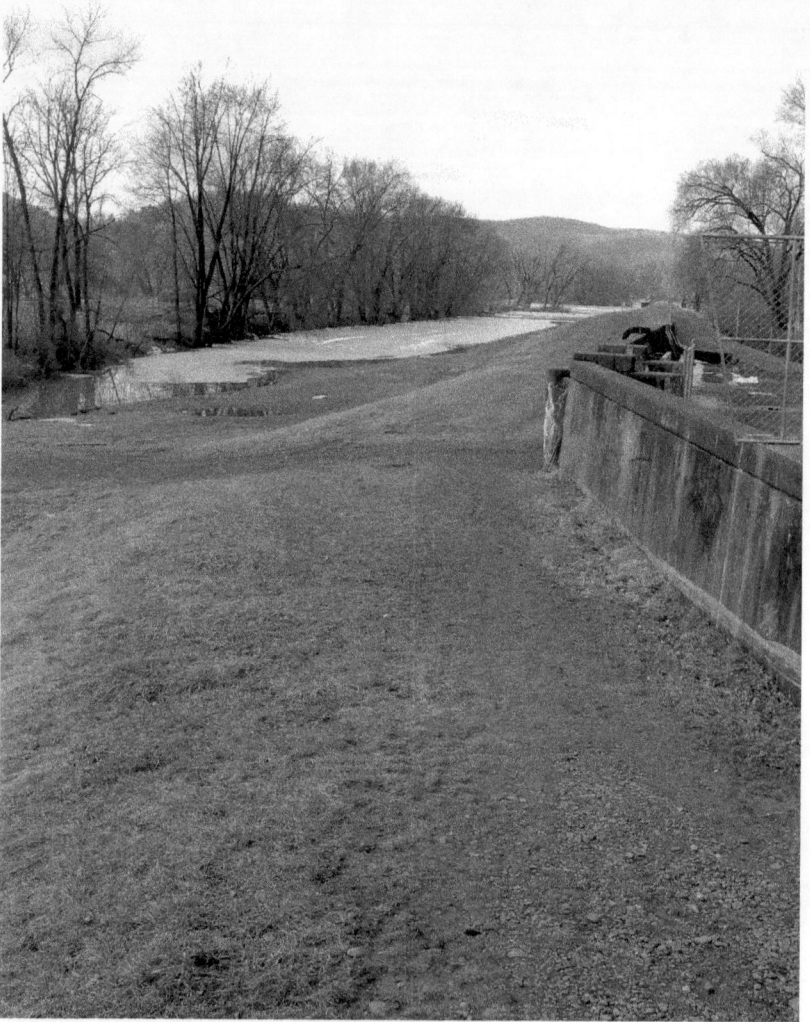

Site of the former Elmira Prison Camp today.

A SOLDIER'S STORY.

I have been requested to write some incidents, experiences and observations of prison life during the war of 1861-'65. After thirty-eight or thirty-nine years it is somewhat difficult to recall anything like all that transpired in those dark days. Some people say it is time to stop talking about that war. Now, that would be a hard thing for those who lived in those days to do: stop talking about the war. The men, women and children at home had almost as hard a time as those at the front—not quite so dangerous, yet it required courage and true patriotism to stand in their places. Furthermore, it seems necessary, in order to keep history straight, that those who lived and participated in that part of our history should occasionally be heard from, otherwise those who write so much, who live north of the Mason and Dixon's line, would make our rising generation believe what is false. So I say to all such: "Nothing in the past is dead to the man who would learn how the present came to be what it is." Much has been written and said by our Northern friends as to the suffering of the Union soldiers in Southern prisons—Andersonville, Salisbury and other places—during that war. They draw an awful picture of their poor soldiers suffering and dying in Southern prisons. In some respects this was true. To be in prison of itself was bad enough, but to be there without proper food or medicine was very bad indeed. The South did not have the means, neither the medicine, but the prisoners in our care were put on the same footing as our own poor soldiers. The question is: Who was to blame for this state of things? The Confederate authorities made proposition after proposition for exchange of prisoners, but the Government at Washington positively declined. It is said that

General Grant said: "It was hard, and a great sacrifice, to leave the Union soldiers in Southern prisons, but it must be made; that the Confederates could not afford to leave their men in prison for want of men to take their place, but the United States could; to exchange the prisoners the Confederates would return to the army and go to fighting again." So here is the key to the responsibility for all the suffering and deaths on both sides in the prisons. The Confederate Government offered to let them send medicine South for their sick prisoners, but they declined to do that. It must be remembered the Confederate Government was shut in from the outside world, and could not secure necessary medicine, etc. Now, as to Andersonville, it was under the command of Wirtz, and since men have had time to cool off it has long since been decided that the hanging of that poor man was simply murder. He did the best he could for the poor prisoners there. General Dick Taylor in his book, "Destruction and Reconstruction," gives the following account of meeting with Wirtz, as his troops were passing Andersonville, during the march of Sherman through Georgia, in 1864: "In this journey through Georgia, at Andersonville, we passed in sight of a large stockade inclosing prisoners of war. The train stopped for a few moments, and there entered the carriage to speak to me a man who said his name was Wirtz, and that he was in charge of the prisoners near by. He complained of the inadequacy of his guard and the want of supplies, as the adjacent country was sterile and thinly populated. He also said that the prisoners were suffering from cold, were destitute of blankets, and that he had not wagons to supply fuel. He showed me duplicates of requisitions and appeals for relief that he had made to different authorities, and these I endorsed in the strongest terms possible, hoping to accomplish some good. I know nothing of this (man) Wirtz, whom I then met for the first and only time, but he appeared to be in earnest in his desire to mitigate the condition of his prisoners. There can be but little doubt that his execution was a 'sop' to the passions of the 'many-headed.' " So, then, poor Wirtz was made a scape-goat to cover the sins of those who could have had those poor prisoners released at any time but would not. The sacrifice was made to quiet the poor prisoners and their friends. Many things will be settled at the great Assize, when the Judge of all shall sit in judgment.

Let us have the official record on prison life, and see the truth of history:

United States prisoners held in Southern prisons, . 270,000
United States prisoners died in Southern prisons, . 22,000
About 8 per cent.
Confederate prisoners held in Northern prisons, . . 220,000
Confederate prisoners died in Northern prisons, . . 26,000
About 12 per cent.

The North is said to be more healthy than the South, and yet of the 270,000 Northern soldiers in Southern prisons, only 22,000 died, while of the 220,000 Confederates in Northern prisons (50,000 less than we had of theirs) 26,000 died. The deaths in Northern prisons exceeded the deaths in Southern prisons four thousand men. While about eight per cent of the Union prisoners died, about twelve per cent of the Southern prisoners in Northern prisons died. "Tell it not in Gath, and publish it not in the streets of Askelon." Facts and figures are wonderful things. Now, I have made this long statement before coming to the "incidents of prison life," as seen by myself *et al.* I have done so for the purpose of trying to keep the record correct, that justice might be done to all, and history speak the truth.

I was shot in the first charge that was made at Spottsylvania Court-House, Virginia, early on the morning of the 9th day of May, 1864. The charge was made by our brigade, composed of the Fifth, Twelfth, Twentieth and Twenty-third N. C. Regiments, led by General R. D. Johnston. The charge was a success so far as the enemy in our front were concerned, but our lines were overlapped by Burnside's troops. Our regiment (the Twelfth) and our company (A), being on the extreme right, were exposed to an enfilading fire clear across an open field; so we were exposed to a fire from front and from the right. The enemy had torn down a rail fence and made temporary breast-works in our front, from which our men drove them, but could not hold the position because Burnside's whole army corps was on hand, and could easily have cut off our little brigade; so General Johnston gave the command to fall back. As our troops fell back, Sergeant Silas Smyre (now county commissioner of Catawba) and Corporal E. G. Bost endeavored to carry me from the battlefield. They were so exhausted from marching and fighting that they could not hold me up so as to prevent the crushed leg from dragging on the ground. To prevent their being captured, I begged them to leave me to my fate. (May I never forget this act of kindness by these brave men, who risked so much for me.) I was in the broiling hot sun, without water, my canteen having been shot in the fight, and the water all run out.

I was concealed from the enemy by some shrubbery. Late in the afternoon I realized that I could not live without water. The loss of blood, together with the burning rays of the sun, made me feel that life was about to ebb out; so I called to the enemy and surrendered. Here I commenced the life of a prisoner, which lasted ten months. Besides the suffering from wounds, the humility, the loss of liberty, the absence of all friends and loved ones, no face but that of enemies, was just about as much as I could bear up under in my condition. In that hour home and friends would have been "a haven of rest" sure enough.

The day following, May 10, 1864, when I was laid on the slaughter table, my eyes caught the sight of arms and legs piled on the ground—an indication of what I might expect. Dr. Cox, of Ohio, examined my leg. The only conversation that passed between us was this: I said, "Doctor, can you save my leg?" He replied, "I fear not, Johnny." Chloroform was applied, and when restored to consciousness I was minus one limb. I lay there in what was designated "a field hospital" for two or three days without any further attention to the wound, and the result was the flies "blowed" the amputated limb, and when I reached Alexandria City, some days later, the nurse who dressed the wound found that I was being eat up by the vermin. Just here I will state that on the last day spent at the field hospital there was a great rush in gathering us up in ambulances. Under great excitement, I said to the doctor who was supervising the movement: "Doctor, what is the matter?" He replied that "Burnside was falling back to get a better position." I had been in the army long enough to know that was an evasive answer. The fact was that our troops were driving Burnside back, and the Federals were not willing to lose any of their prisoners though maimed for life. The roads from this place were cut to pieces by the artillery and wagon trains of the Union army going to the front. Those of us who were badly wounded cried for mercy. No mercy came until we reached the boat-landing, where we (those living) were transferred from ambulance to the boat. I do not know how many died en route from the battlefield to the boat-landing. I do know that Charles P. Powell, Adjutant of the Twenty-third North Carolina Regiment, who had lost his leg just as I had, died on this trip, and they stopped on the roadside and covered him up. This young man Powell was from Richmond County, N. C. He was a private soldier at Malvern Hill, July, 1862. When in line of battle, in front of the artillery, a shell fell in the ranks. The men could not leave the line of battle. There lay the shell, sputtering, ready to explode. Young Powell sprang up, grappled the shell and "soused" it into a pool of water near by. What a risk was that! Yet that heroic act may have saved the lives of several men. Later that day he was wounded, and again at the battle of Gettysburg in July, 1863, and died as above stated. On page 189 of Volume II, North Carolina Regimental Histories, it is stated that C. P. Powell, Adjutant, was killed on the 9th of May, 1864, whereas the truth is he was shot on the 9th and his leg was amputated, and about the 11th or 12th of May he was jolted to death between Spottsylvania Court-House and Bell Plains. I venture the assertion that he was not buried two and a half feet deep; and the place is unknown to his people, who think he was buried on the battlefield. We were shipped to Alexandria City, where I spent three months in the "Marshall House," where the proprietor, Jackson, shot and killed Colonel Ellsworth, who tore down his Confederate flag in April, 1861, and Jackson was killed by Frank Brownwell, of Colonel Ellsworth's regiment. This hotel was used as a prison hospital for those who were permanently disabled. For awhile the patriotic women of Alexandria were permitted to visit us, and often when they would bid us good-bye a "green-back" bill or something else was left in our hand. However, before we were removed from there the good women were prohibited from coming to see us.

While a prisoner here our troops, under General Early came down near Washington City, and there was great excitement in Washington and Alexandria, for it did seem that the Confederates were going into Washington. We prisoners were expecting to be released and get home, but our expectations were soon blasted by the Confederates having to retreat back to the south side of the Potomac, and did not come *via* Alexandria. My next move was to the Lincoln Hospital in Washington City. Here I spent about two months. After I could walk with crutches I was transferred to the old Capitol Prison. I was honored with a seat in the old Capitol, but had to look through iron bars. While here I was guilty of "cruelty to bugs," if not to animals, in the common acceptation of that term. (Just here by way of parenthesis.) I know how to appreciate the traveling man's experience given by "Red Buck," in *Charlotte Observer,* of September 11, 1903. Night after night I suffered from the onslaughts of those "bugs"—no telling how much I endured. "Weeping endureth for the night, but joy cometh in the morning." They had all the "innings" at night, but in the morning I would take my turn at the bat. As soon as it was light enough to see I would sit upon my humble couch (I was myself a picture of humility) and commence a war of revenge. As they would take to the wall I would go for them, and before I left that prison many, many "bugs" were slaughtered, as the blood-stained wall bore testimony. Yes, that wall was well striped with Confederate blood. The loss of blood in that way, if not with as much pain, was attended with much more genuine disgust. How much I would have liked to "express myself," but my lips were hermetically sealed. I learned how to sympathize with Pharaoh and his people, though there is no statement that any of this kind were sent on him when Moses and the Israelites were asking permission to leave. In November, 1864, I (with others) was shipped off to Elmira, N. Y. "Thus when I shun Scylla, your father, I fall into Charybdis, your mother."

Leaving the old Capitol Prison, I got away at least from the multitude of B. B.'s, but I ran into the B. L.'s—army body lice, or what the soldiers call "grey backs." Later on I may speak of my experience with this pest while in the small-pox camp.

We reached Elmira, N. Y., on Sunday morning. Being in the mountains, the ground was covered with snow. Arriving at the barracks, we were lined up (I was on my crutches, · and had to stand there on one foot for what seemed to me a very long time) just inside the gate, negro soldiers on guard. The commanding officer, Major Beal, greeted us with the most bitter oaths that I ever heard. He swore that he was going to send us out and have us shot; said he had no room for us, and that we (meaning the Confederate soldiers) had no mercy on their colored soldiers or prisoners. He was half drunk, and I was not sure but that we might be dealt with then and there. Then we were searched and robbed of knives, cash, etc., and sent into various wards. While we were standing in the snow, hearing the abuse of Major Beal, some poor ragged Confederate prisoners were marched by with what was

designated as barrel shirts, with the word "thief" written in
large letters pasted on the back of each barrel, and a squad
of little drummer boys following beating the drums. The
mode of wearing the barrel shirts was to take an ordinary
flour barrel, cut a hole through the bottom large enough for
the head to go through, with arm-holes on the right and left,
through which the arms were to be placed. This was put on the
poor fellow, resting on his shoulders, his head and arms coming through as indicated above; thus they were made to march
around for so many hours and so many days. Now, what
do you suppose they had stolen? Why, something to eat.
Yes, they had stolen cabbage leaves and other things from
slop barrels, which was a violation of the rules of the prison.
One large, robust prisoner from Virginia was brought into
the surgical ward where I was, having been seriously wounded
by one of the guards. On inquiry, I learned that the poor
fellow was caught fishing out scraps from a slop barrel and
was shot for it. A small, very thin piece of light-bread with a
tin pint cup full of what purported to be soup twice a day
was the rations for the prisoners. I heard the men say: "My
soup has only three eyes on it"—meaning there was no grease
in it—only hot water. Now, this fare was not enough to
sustain life in healthy, able-bodied men. The result was that
where they could not make something—make rings, etc.—and
thus secure something from the sutlers, many, yea hundreds
of the poor fellows would be attacked with dysentery—so
common and often so fatal in camp, and especially in prison
life. The food they had seemed to be only enough to feed
the disease; the result was that scores and hundreds died.
Speaking of the light-bread, the Confederates would sometimes hold it up and declare "that it was so thin that they
could read the *New York Herald* through it"; then they
would grab it and squeeze it up in one hand till it looked
about like a small biscuit. Men died there for the want of
food. I do not know, it may be that the Government issued
enough rations, but it had to pass through too many hands

before reaching th soldiers. The truth is that there was a great deal of speculation and swindling carried on in the prisons; and I am ashamed to say it, yet it is true that sometimes some of our own men were engaged in the conspiracy to cheat and defraud their fellow-prisoners. It was in this way: those in charge of the prison would take Confederates and make ward-masters, etc., of them (like in prisons now a few are made "trusties") ; and a little authority, even of that kind, would ruin some men. Some prisoners, like Jeshrun, grew fat, but others starved for want of suitable food and enough of it. Well, to go back a little, while standing there, receiving the profane blessing from Major Beal, I saw drawing near as he dared to venture an old fellow-prisoner that I had met in Washington, who had preceded me to this place. I do not remember his name. I had at Washington nicknamed him "Softy." He recognized me, and as Beal closed his eloquent abuse, and we were ordered to march into the barracks, "Softy" ventured in a low tone to speak to me. His greeting was: "Sherrill, you have come to hell at last. Did you see those four-horse wagons going out? They were full of dead men, who died last night. They are dying by hundreds here with small-pox and other diseases." He was discovered by one of the guards (standing too near us). He hollowed at him: "Get away from there." He got away immediately, if not sooner. When I reflected on the situation—the cursing major, the colored guards, the robbing us of our little stock of valuables, the barrel shirts, the wagons with the dead, the appearance of some of the living, the earth covered with snow— I thought, "Well, 'Softy' has given a true bill." When I was located, I found I had kinsfolk there: J. U. Long (now chairman of the board of county commissioners), Nicholas Sherrill and W. P. Sherrill. There may have been others, but I do not recall them now. My haversack had been supplied with rations on leaving Washington. When I was located in the ward, "Nick" Sherrill came to see me. Of course we were glad to see each other, for it had been many moons since we had met. We were not in the same command in the army. "Nick" asked me if I had anything to eat. I replied, "Yes." He said: "I want to trade you a cup, spoon, etc., for some bread; I am about perished." Poor fellow, he looked the picture of despair. I said: "Nick, I do not want your cup and spoons, but you are welcome to what I have." He devoured in short order all that I had, and wanted more. Poor fellow, he soon died, as did W. P. Sherrill; died away from home and loved ones, buried by their enemies. I had to spend several days in the barracks before I was transferred to the surgical or hospital ward. I was there long enough to know why Cousin Nicholas was so anxious for my bread. After I was placed in the surgical ward of the hospital I fared fairly well—a great improvement over the fare out in the wards of the regular prison. After a few weeks I was taken with small-pox, and of course was transferred over S. Creek to the small-pox camp. I was carried over on a cot, or "stretcher," with blanket thrown over my face. When I reached the place, and the blanket was removed, I found myself in a large "wall tent," with several cots, or "bunks," about two and a half feet wide, with two Confederates on each "bunk," in reverse order, i. e., A's head at one end and B's at the other— so your bed-fellow's feet were in very close proximity to your face. They were all sandwiched in this way, because the bed was too narrow to admit of the two to lay shoulder to shoulder. On waking up on a morning one of these poor fellows would be dead and the other alive; this, of course, occurred day after day, and night after night. Well might those poor fellows, who had spent at least a part of the night with a corpse for a bed-fellow, have exclaimed with St. Paul, "Who shall deliver me from the body of this death?" When I took in the situation, I told the man who was going to place me on a bunk by the side of a poor fellow bad off with that awful disease (and who finally died) "that he could not put me on there." He replied "that he would show me whether he could or not." I stuck to it that I would not be put there. The

16

fellow went and brought in the ward-master, and when he appeared it was Jack Redman, from Cleveland County, Company E, my regiment. Redman said, "Why, hello, Sherrill, was it you that was raising such a racket?" I told him it was. He wanted to know what was the matter. I explained that with my amputated limb it would never do to put me on a bunk with another fellow, and he finally consented to arrange for me to have one to myself. I said: "Redman, you must grant me another favor." He wished to know what it was. I replied: "I want you to let me keep my blanket that came over from the surgical ward." "Why so, Sherrill?" I said: "Jack, you see those blankets that you fellows have been using on these men—there are five 'army lice' to every hair on the blanket." Redman took a hearty laugh. He knew there was more truth in it than poetry, so he granted my request. Redman had had small-pox and was an "immune," hence was made a ward-master. He was especially kind and considerate towards me. When I got well and was carried away, I never knew what became of him. Some of our men who felt that the thing was gone, and that we could not succeed, never came back South. I am inclined to think that Redman did that thing. After the doctor had declared me well, and directed that I should be removed back to the hospital ward from whence I came, this was indeed glorious news; for of all the diseases that flesh is heir to, small-pox is the filthiest. The small-pox such as we had there was "sure enough" small-pox. Such as we have in North Carolina these days, in comparison with that, is only make-believe. I don't think it an exaggeration to say that seven out of ten who had it died. I was carried over into what was called a bath-house, where I was placed in a large bath-tub of water, almost too hot to bear. The Yankee soldier who had charge went out to look after something else or to loiter around, and I waited and waited for his return (the water was beginning to get cold) so I could get out and get clothing to put on. The atmosphere of the room was colder, if anything, than the water. I was in great distress, and it seemed that I could make no one hear me; so I had to wait the return of the villain, who finally came when the water in the bath-tub seemed to me to be nearly to the freezing point. He came, bringing a full Yankee suit, and when I gave him a piece of my mind he apologized and begged me not to speak of it—said he had actually forgotten me. When I reached the hospital ward I was a blue man in feelings and in appearance. I was dressed in a Yankee suit, even to a cap. I felt humiliated, and my skin was blue from cold. But for the kindness of my comrades there, giving me of their allowance of spirits that night, I don't know but what I would have gone hence.

Along toward the close of February, 1865, I with others, was marched to the train and shipped to Richmond. I think that was the happiest day that I ever experienced in my life. To get out of that death-hole was enough to make one happy; and to add to it the prospect of getting home to friends and loved ones, from whom I had been so long separated, not having heard from them in ten months, was indeed a treat. Many and great changes had taken place since I had left Dixie. I never did doubt that we would eventually succeed. I presume I was cheered up and was kept optimistic from the many rumors all the time in circulation that France and England would soon recognize our independence; which, of course, never took place. The air was filled with that and other rumors, not only in the Confederate army, but even in prison. Such rumors of great victories for the Confederate arms were all the time circulating among the poor fellows. As I came on from New York it looked to me as if the whole world was being uniformed in blue and moving toward General Grant's army. As we came up the James River, both sides were lined with soldiers dressed in blue. When we came to the Confederate lines, seeing such few ragged men confronting all that blue host, my courage came near failing me. In fact, I could not see how this little thin line of Confederates could hold at bay such a multitude of well-fed, well-equipped men. The patriotic women of Richmond tried to be cheerful, but I could see plainly enough that they were depressed. While they were just as kind in their attention to the returning soldiers as in former days, yet it was evident that the cheerful hope of former days was gone. When I reached home I soon learned that many who were living on the 9th of May, 1864, when we made that charge, had been numbered with the dead. Among others was my nephew, James Ferdinand Robinson, a young man a few months younger than myself, a great favorite in the company, full of humor and wit. He was a sharp-shooter, and was found dead on the 12th of May, 1864, by Frank Turbyfield, of the Twenty-third Regiment. After the fighting on the morning of the 9th, he wrote a letter in pencil to his father, Marion Robinson, in which he stated: "My Uncle Miles was killed in the charge made early this morning." Two days later he was killed. I got home to read his letter relative to my death; but he, poor fellow, was gone. I have not seen the letter since 1865; so I only quote from memory what I remember.

Such is war. Many people have an erroneous idea about that war. They blame President Davis and President Lincoln for the whole thing; when in fact they were only placed at the head. Both made blunders; so would any one else in their positions. Davis was not an original secessionist, but went with his State. He was a United States Senator at the time, from Mississippi. He had served with distinction in the war with Mexico. Who has not read of "Colonel Jeff. Davis and his brave Mississippi riflemen"? Mr. Davis did not desire to be President; he desired to go in the army. He had been Secretary of War of the United States; had, as stated above, served in the United States army; so it was natural for him to prefer the army to being President. As to his taking the responsibility of making peace sooner, I have seen it stated that had he attempted to do so in 1864, on any terms save independence, the army and the people of the South would not have submitted to it. I think myself this is true. He, as well as General Lee, had a hard time; they were both weighed down with trouble, cares and responsibilities. He had no more to do with the assassination of President Lincoln than you or I. He was cast into prison, manacled and placed in a dungeon. (General Miles would be glad now if he never had put shackles on him.) A soldier was placed where an eye always rested on Mr. Davis. This was a great annoyance to him.

General Dick Taylor, who succeeded in getting permission from President Johnson to visit President Davis at Fortress Monroe, makes the following statement: "It was with some emotion that I reached the casement in which Mr. Davis was confined. There were two rooms, in the outer of which, near the entrance, stood a sentinel, and in the inner was Jefferson Davis. We met in silence, with grasp of hands. Afterwards he said: 'This is kind, but no more than I expected of you.' Pallid, worn, gray, bent, feeble, suffering from inflammation of the eyes, he was a painful sight to a friend. He uttered no plaint, and made no allusion to the irons. He said the light kept all night in his room hurt his eyes, and the noise made every two hours by relieving the sentry prevented much sleep; but that matters had changed for the better since the arrival of General Burton, who was all kindness, and strained his orders to the utmost in his behalf,' etc." Mr. Davis was no doubt a great and good man, for General Taylor, on speaking of some kindness shown to him during the war, said: "No wonder that all who enjoy the friendship of Jefferson Davis love him as Jonathan did David." Had Mr. Davis been a traitor and rebel any more than other leaders of the South, and had he been guilty as charged, of course he would have been tried and executed. It was not done simply because it would have been an open violation of law, and the people of our country had had time to cool off. So Mr. Davis was released. We all believe that had Mr. Lincoln lived we never would have had to go through the farce and humility of reconstruction. Excuse me, Mr. Editor, for this divergence. I have done so "lest we forget; lest we forget." There are many humorous, ludicrous, laughable things that occurred in prison life, connected with the negro soldiers (sparring between the colored guard and the Confederate prisoners) that will not do to publish; so I forbear to give any of them.

It is indeed wonderful how the prisoners would work to make a little money. One of the most common occupations was to make finger rings; they did some real nice work. Some of the men would secure a few cents, and on that little capital build up quite a business. Some had teachers and attended school. The teachers were, of course, fellow-prisoners with the pupils. As before stated, I was in the surgical ward while in New York, and had no personal experience in the traffic and trading above alluded to, for it was not allowed in the hospital wards. Mr. John Gray Bynum, of Mountain Creek township, was a ward-master while a prisoner at Elmira (and made a good one, too). He could give some rich incidents of prison life; and so could our mutual friend Phillip A. Hoyle, who was a prisoner at Point Lookout, Md. It may not be generally known that Mr. A. A. Shuford, of Hickory, one of our successful business men, made his start as a trader while a prisoner of war. It is my understanding that such is the case. It was while in prison that Mr. Shuford manifested a talent and a liking for trade and traffic, and on a small scale made a success while in prison. Having thus imbibed the business spirit while in prison, on his liberation and return home he left the farm and old homestead and went to Hickory and engaged in business with his brother "Dolph" and W. H. Ellis. How well he has succeeded is a matter of history, and who can tell what influence his experience in prison may have had on his subsequent life? A. A. Shuford and P. A. Hoyle belonged to the gallant Twenty-third North Carolina Regiment and suffered together at Point Lookout, where the water was impregnated with copperas, thus causing the death of thousands of as brave men as ever carried a gun. I am reminded that General Lee says in his memoirs that he used every effort and means at his command to effect an exchange of prisoners, but General Grant refused.

As before stated, General Grant refused to exchange as a war measure, and it had the desired effect.

That there were some men in uniforms who might be classed as brutes is not to be denied; we are thankful the number was comparatively small. In the campaign into Maryland in 1862, our regiment was in the division commanded by the gallant Gen. D. H. Hill, who held the mountain passes against overwhelming numbers. My younger brother, James Albert Sherrill, who had been with us only six months, fell dangerously wounded just at the time the command was given to fall back. Of course he fell into the hands of the enemy; there, lying weltering in his blood, the enemy came on him, and instead of ministering to his wants, a brute in human form in uniform took his bayonet and stabbed the poor boy to death. I did not see this, but Alfred Sigmon, of Catawba County, who was also wounded, was an eyewitness to the tragedy. I give this incident as it came near to me; many others just as cruel might be given. It would not do to hold General McClelland or his true soldiers responsible for the conduct of a drunken, cowardly brute. The Union army was afflicted by having foreign soldiers who could not speak the English language. We have met the Union soldiers when many of them were so drunk they could hardly tell what they were doing.

There never was any trouble between true soldiers, whether they wore the blue or the gray. It was the warlike civilians who did not fight and the soldiers who were mere hangers-on and camp followers that made the trouble. But for the influence of General Grant and other army officers we would have fared much worse in the South after the close of the war than we did; they, as conquerors, became our protectors. The true soldiers could be seen exchanging coffee for tobacco, going in bathing at the same time, in the same river; and when the enemy fell into his hands as a prisoner he would empty his own haversack and the canteen to relieve his prisoner. When there was no fighting going on, the soldiers of the two armies were on the best of terms. The outrages committed on either side during the war were not attributable to the true soldier; neither can the outrages perpetrated on the South after the war be charged up to the United States Army proper, but to the "bummers," who were no good in the army or at home.

The storm has long since gone by. The true soldier has no prejudice against the soldier who fought on the other side. The blue and the gray have since worn the blue in the war with Spain—an evidence of reconciliation between the Confederate and Union soldiers of 1861-'65.

Since writing the foregoing sketch I have received the following "Memorial Day Ode," from the pen of my friend, Rev. G. R. Rood, preacher in charge of Millbrook Circuit. It is so appropriate I let it be the closing chapter:

MEMORIAL DAY ODE.

The past is dead, long live the past;
And may its memory ever last
In hearts through which the Southern blood
Leaps on its way an untamed flood.
For we who bear the Southern name
Look on the past and find no shame
Attached to the cause which, though lost,
Was worth the life-blood which it cost.
And though the mournful willows wave
Over the low mounds which we lave
With bitter tears, we feel,
We know the future will reveal
That each martyred hero doth wear
A crown of heavenly laurel fair.
Each spot which heard the dying moans,
And which in death received the bones
Of those who freely gave their all,
In answer to the Southland's call—
No matter where they may be found,
Such spots are sacred, holy ground.
The heroes who sleep 'neath the sods
Rest in sweet peace, their souls are God's,
Until the Judgment trump be blown,
And wrong forever is o'erthrown;
Then they will rise up one and all
To answer to the Last Roll Call.

G. R. ROOD.

Millbrook, N. C.,
        May 7, 1904.

24

About Miles Osborn Sherrill

Sherrill was born in 1841 at Sherrills Ford, in Catawba County, North Carolina. According to his service records he joined Company A, 12th North Carolina Infantry on April 27, 1861.

He was shot at Spotsylvania Court House, Virginia and his fractured leg was amputated in May 1864 at the age of nineteen. After spending several months in the Lincoln General Hospital and the Old Capitol Prison in Washington, DC, he was sent to Elmira in December 1864.

On February 9, 1865 he was "transfered for exchange" to James River. He received orders to return home in Catawba County on February 25, 1865.

He died on April 8. 1919 and is buried in Green Hill Cemetery in Greensboro, North Carolina.

About the Elmira Prison Camp

Where:
Elmira, New York - a city on the Chemung River

What:
At the beginning of the Civil War, Elmira had been a military recruiting depot where soldiers attended basic training. Later in the war Elmira was chosen as a draft rendezvous, and then a new prisoner of war camp. The first prisoners arrived at the camp on July 6, 1864. The last prisoners left the camp on July 11, 1865.

Who:
12,122 Confederate enlisted and non-commissioned officers POWs were assigned to Elmira.

When:
July 6, 1864 - July 11, 1865

Why:
Elmira had ample barracks at the time. The North needed a place to house prisoners.

What went wrong:
Barrack space was ample for 5,000 prisoners, but 10,000 arrived and were forced to live in tents along the Chemung River. Keep in mind the weather in New York State from October to April. Lack of nourishing food, extreme bouts of dysentery, typhoid, pneumonia, smallpox, lack of doctors and medicine, and flooding of the Chemung River, caused the deaths of 2,963 prisoners who are interned in Woodlawn National Cemetery on Elmira's northside. Forty-eight more who died in the Shohola Train Wreck while en route to the prison are also buried there.

Political Views - Then & Now: In the time of the prison camp in Elmira, the North was right and the South was wrong and the prisoners were (mis)treated accordingly. For over 130 years the mistreatments of prisoners were dismissed as rumor.

> "The horrors of a camp where prisoners of war are crowded into a confined space, poorly clad, uncomfortably housed, insufficiently fed, and scantily provided with medical attendance, hospital accommodations, and other provisions for the sick, form one of the most deplorable features of any war, but none of these can apply with truth to the camp at Elmira, nor can they be attached for a moment to the reputation or become a portion of the history of the fair valley of the Chemung."
>
> - *The History of Chemung County*, Ausburn Towner, 1892.

History books of the time held the denial and heralded the excellent care of the Southern prisoners of war. The truth about the camp (the lack of food, medicine, and shelter) finally began surfacing in the past few years with several new books about the Elmira Prison Camp. It had taken over 130 years to admit the abuse.

A Memorial Day display at the Woodlawn National Cemetery.

The Confederate Monument at the Woodlawn National Cemetery.

More Civil War titles from
New York History Review

Diary of a Tar Heel Confederate
by L. Leon

The Elmira Prison Camp
by Clay Holmes & Diane Janowski

In Their Honor: Soldiers of the Confederacy -
The Elmira Prison Camp
by Diane Janowski

www.ingramcontent.com/pod-product-compliance
Lightning Source LLC
Chambersburg PA
CBHW030011040426
42337CB00012BA/743